MATHadazzles

Mind Stretch Puzzles

Reasoning with Integers Volume 3

Authors

Carole Greenes

Mary Cavanagh

Contributors

Renee Ashlock, Channing Bogle, Nancy Foote, Kristi Larson,
Katie MacDonald, Holly Natonick, Jane Placencia,
Theresa Thivierge, Jen Tom, Diana Volanti, Cara Wright, Michael Xu

Editors
Javy Duarte, Senior Editor
James Kim
Jason Luc
Lesley Le
Tanner Wolfram

Cover Design
Yifan Tian

MATHadazzles *Mind Stretch Puzzles*
Reasoning with Numbers Volume III

MATHadazzles are number puzzles that will develop your logical reasoning abilities, your sense of numbers (different types of numbers, their characteristics, and operations with them), and your persistence in solving problems. Once you start, you won't be able to stop UNTIL you successfully solve all of the puzzles!

What is an Integer MATHadazzle? An Integer MATHadazzle is a 3-by-3 grid with circles at the end of each row and column. Some grid cells have clues about the numbers that will fill those cells. The numbers in the circles at the ends of the rows and columns are the row and column sums.

What's Your Job? Based on clues provided in some of the grid cells, you place the numbers -5, -4, -3, -2, -1, +1, +2, +3, and +4 in the nine cells, so that the row and column numbers add up!

What are The Clues? Clues describe types of numbers, characteristics of those numbers, or results of operations with the numbers. There are 10 types of clues. These may appear singly or in combination. Some clues will apply to both positive and negative numbers. Some clues apply to only positive numbers.

Clues that apply to positive numbers only:

Even: Number that is divisible by 2. The even numbers in the set of integers -5 to +4 are 2 and 4.

Odd: Number that is not even. The odd numbers in the set of integers -5 to +4 are 1 and 3.

Multiple: Number that is the product of a counting number and a whole number. Examples: 8 is a multiple of 4, because 2 x 4 = 8, and 6 is a multiple of 1 because 6 x 1 = 6.

Factor: One of two or more numbers that produce a product. Examples: 2 and 3 are factors of 6 because 2 x 3 = 6, and 2 and 4 are factors of 8, because 2 x 4 = 8.

Prime: Number divisible by only two numbers, itself and 1. Important: Itself and 1 must be different numbers. Examples: 3 is only divisible by 1 and 3. 7 is only divisible by 1 and 7.

Composite: Number greater than 1 that is not a prime number. That is, it has more than two different factors. Example: 4 can be divided by 1, 2 and 4. 8 can be divided by 1, 2, 4 and 8.

Clues that apply to both positive and negative numbers:

Square Number: Number that is the product of a non-zero number and itself. Examples: 4 is a square number because 2 x 2 = 4 and -2 x -2 = 4. 9 is a square number because 3 x 3 = 9 and -3 x -3 = 9.

Square Root of a Number: Number that when multiplied by itself has a product equal to a given number. The square root of 16 is 4, because +4 x +4 = 16 and -4 x -4 = 16. The square root symbol is $\sqrt{}$. Example: $\sqrt{16}$ = 4, or -4.

Exponent: A small number to the right and above a base number that indicates the number of times that base number should be used as a factor in multiplication. In this book, all exponents are positive, although the base number can be negative. Example: 2^3 = 2 x 2 x 2 = 8. In this example, 2 is the base number, 3 is the exponent, and 8 is the product. Another example: -5^3 = -5 x -5 x -5 = -125.

Computation Operations: All four operations with integers are incorporated into the clues. These are: addition, subtraction, multiplication, and division. When several different operations appear in a number sentence, then the order in which they are performed follow the Fundamental Order of Operations: Parentheses (all operations within parentheses are computed first), Powers (all base numbers with exponents), Multiplication and Division (from left to right), and finally, Addition and Subtraction (from left to right).

In Volume III, you will find 78 *MATHadazzles* using integers. If you are interested in improving your dazzling solution talents, consider getting Volumes I and II, with more puzzles, but using the counting numbers 1 – 9 only. Answers are at the back of each book. Look for more *MATHadazzles* coming soon!

Enjoy Solving!

Your MATHadazzling Authors, Contributors and Editors

1

Put these numbers in the squares   -5 -4 -3 -2 -1   1   2   3   4

Add across →

Add down ↓

Sums are in ◯

1^{100}			**6**
		1×-2	**1**
$9 \div -3$		$30 - (5 \times 7)$	**-12**
2	**-3**	**-4**	

Put these numbers in the squares   -5 -4 -3 -2 -1   1 2 3 4

Add across ⟶

Add down ↓

Sums are in ◯

Positive		Odd prime	⟶ **3**
	$4^3 \div 4^2$		⟶ **-3**
	Prime	$6 \div -2$	⟶ **-5**

-5   **5**   **-5**

3

Put these numbers in the squares   -5 -4 -3 -2 -1   1   2   3   4

Add across ⟶

Add down ↓

Sums are in ◯

	2 - 4	-6 ÷ 2	
			-6
Odd			**-8**
Square greater than 1		Even prime	**9**

4   **-3**   **-6**

Put these numbers in the squares   -5 -4 -3 -2 -1   1   2   3   4

Add across ⟶

Add down ↓

Sums are in ◯

$\sqrt{25}$		$\sqrt{1}$	**-4**
2^2	$\sqrt{16}$		**3**
	1^2		**-4**
-3	**-1**	**-1**	

5

Put these numbers in the squares   -5   -4   -3   -2   -1   1   2   3   4

Add across $\longrightarrow$

Add down $\downarrow$

Sums are in $\bigcirc$

$\sqrt{16}$	Prime	Even prime	
			$\boxed{9}$
	$\sqrt{9}$	$16 \div -8$	$\boxed{-9}$
$\sqrt{25}$			$\boxed{-5}$

$\boxed{-5}$ $\boxed{-1}$ $\boxed{1}$

6

Put these numbers in the squares   -5 -4 -3 -2 -1   1   2   3   4

Add across ⟶

Add down ↓

Sums are in ◯

			-3
	$1^8 \times 2^1$	$3^2 - 3^1 - 3^1$	**6**
$5 \div \text{-}5$		$\text{-}4 - 1$	**-8**
-3	**4**	**-6**	

7

Put these numbers in the squares   -5 -4 -3 -2 -1 1 2 3 4

Add across →

Add down ↓

Sums are in ◯

$6 - 11$		$8 - 9$	**-5**
			-4
	$8^{0} \times 3$	-3×1	**4**
1	**0**	**-6**	

8

Put these numbers in the squares   -5 -4 -3 -2 -1   1 2 3 4

Add across ⟶

Add down ↓

Sums are in ◯

	-4×9^0		
			-6
$\sqrt{4}$			
			-3
Square		18^0	
			4
-3	**-3**	**1**	

9

Put these numbers in the squares   -5 -4 -3 -2 -1 1 2 3 4

Add across ⟶

Add down ↓

Sums are in ◯

	$(3^2 \div 9)^0$		
			-2
$(4-2)^2$		$\sqrt{9}$	
			3
		$1-4$	
			-6

    -2   **-5**   **2**

10

Put these numbers in the squares   -5 -4 -3 -2 -1   1 2 3 4

Add across ⟶

Add down ↓

Sums are in ◯

$8^0 \times 1^0$			-1
	$3 - 6$		-2
$12 - 16$		$\sqrt{16}$	-2
-4	-10	9	

Put these numbers in the squares   -5 -4 -3 -2 -1   1   2   3   4

Add across →

Add down ↓

Sums are in ◯

$3 - 5^0$	$16 \div \text{-}4$		**-1**
			1
$9 \div 3$		$\sqrt{25}$	**-5**
3	**-3**	**-5**	

Put these numbers in the squares   -5 -4 -3 -2 -1   1 2 3 4

Add across ⟶

Add down ↓

Sums are in ◯

Even prime		Factor of all numbers	⟨-2⟩
Power of 3			⟨-2⟩
	2× -2		⟨-1⟩
⟨9⟩	⟨-12⟩	⟨-2⟩	

13

Put these numbers in the squares   -5 -4 -3 -2 -1   1   2   3   4

Add across $\longrightarrow$

Add down $\downarrow$

Sums are in $\bigcirc$

		$8^2 \div 4^2 \times -1$	$\bigcirc$ -4
$12 \div 6$			$\bigcirc$ -5
	-1^3	Positive multiple of 4	$\bigcirc$ 4

$\bigcirc$ 6 $\bigcirc$ -6 $\bigcirc$ -5

14

Put these numbers in the squares   -5 -4 -3 -2 -1   1   2   3   4

Add across →

Add down ↓

Sums are in ◯

			-6
Factor of all numbers	$-45 \div 9$		**-8**
Even prime		Even	**9**
2	**-4**	**-3**	

15

Put these numbers in the squares   -5 -4 -3 -2 -1   1   2   3   4

Add across →

Add down ↓

Sums are in ◯

			-10
	$(5 \div 5)^1$	2^2	2
$6 \div -3$		Odd prime	3
-9	2	2	

Put these numbers in the squares   -5 -4 -3 -2 -1   1   2   3   4

Add across ⟶

Add down ↓

Sums are in ◯

		$3^2 - 12 + 2$
$15 - 8 - 10$		
	$5^2 \div -5$	Factor of all numbers

Circles (right side, top to bottom): **0**, **-5**, **0**

Circles (bottom, left to right): **4**, **-5**, **-4**

Put these numbers in the squares   -5 -4 -3 -2 -1   1   2   3   4

Add across →

Add down ↓

Sums are in ◯

Positive multiple of 4		$32 \div 16 - 1$	③
			-10
$12 \div 6$		Prime	②

① -6 ⓪

18

Put these numbers in the squares   -5 -4 -3 -2 -1   1 2 3 4

Add across ⟶

Add down ↓

Sums are in ◯

		$-252 \div 126$	
			-7
	Neither prime or composite		
			-7
	2^2	Even prime	
			9

◯ **-1**   ◯ **1**   ◯ **-5**

Put these numbers in the squares -5 -4 -3 -2 -1 1 2 3 4

Add across $\longrightarrow$

Add down $\downarrow$

Sums are in $\bigcirc$

$20 \div (2^2 + 1^1)$		Positive factor of 17	$\mathbf{0}$
			$\mathbf{0}$
	$16 \div (-8 - {}^-4)$	Factor of 10	**-5**
$\mathbf{0}$	**-6**	$\mathbf{1}$	

Put these numbers in the squares   -5 -4 -3 -2 -1   1 2 3 4

Add across ⟶

Add down ↓

Sums are in ◯

Composite			**-4**
$(2^3 \div 2) \div 2$	4^0		**6**
	$\sqrt{16} + \text{-8}$		**-7**
4	**-8**	**-1**	

Put these numbers in the squares   -5 -4 -3 -2 -1   1   2   3   4

Add across →

Add down ↓

Sums are in ◯

6^0		$\sqrt{9}$	**2**
		$\sqrt{16}$	**-12**
$\sqrt{16} \div 4^0$		Prime	**5**
2	**-8**	**1**	

22

Put these numbers in the squares   -5 -4 -3 -2 -1   1 2 3 4

Add across ➝

Add down ↓

Sums are in ◯

10^0		$\sqrt{16}$
Prime		-(-3 × -1)
$\sqrt{9}$		

Circles (right side, top to bottom): **0**, **-2**, **-3**

Circles (bottom, left to right): **6**, **-10**, **-1**

23

Put these numbers in the squares   -5 -4 -3 -2 -1   1   2   3   4

Add across $\longrightarrow$

Add down $\downarrow$

Sums are in ◯

$\sqrt{9}$	$\sqrt{16}$	$\sqrt{25}$	**-12**
Composite			**7**
		$16 \div (-4 \times 2)$	**0**
4	**-4**	**-5**	

Put these numbers in the squares   -5 -4 -3 -2 -1   1   2   3   4

Add across →

Add down ↓

Sums are in ◯

$24 \div -8$		Even
Positive	$-(2^0 \div 2^0)$	
		Composite

Circles (right side, top to bottom): **-6**, **3**, **-2**

Circles (bottom, left to right): **-6**, **-8**, **9**

25

Put these numbers in the squares   -5 -4 -3 -2 -1   1   2   3   4

Add across ⟶

Add down ↓

Sums are in ◯

	Composite	Even
Odd		Odd
	$\sqrt{25}$	$\sqrt[4]{16}$

Across sums: ② ① ⑧(-8)

Down sums: (-4) (-4) (3)

26

Put these numbers in the squares   -5 -4 -3 -2 -1   1 2 3 4

Add across ⟶

Add down ↓

Sums are in ◯

$-14 \div 7$	Positive multiple of 2		⟶ **-5**
Positive multiple of 2			⟶ **2**
	$\sqrt{16}$		⟶ **-2**
5	**-1**	**-9**	

27

Put these numbers in the squares   -5 -4 -3 -2 -1   1 2 3 4

Add across $\longrightarrow$

Add down $\downarrow$

Sums are in $\bigcirc$

		$-27 + 26$
$13 \div 13$		$-15 \div 5$
	2^1	

Circles on right: **-7**, **2**, **0**

Circles on bottom: **0**, **4**, **-9**

28

Put these numbers in the squares   -5   -4   -3   -2   -1   1   2   3   4

Add across $\longrightarrow$

Add down $\downarrow$

Sums are in $\bigcirc$

Odd			$\left(\,-5\,\right)$
	Even prime		$\left(\,-5\,\right)$
Even		Odd prime	$\left(\,5\,\right)$
$\left(\,1\,\right)$	$\left(\,-1\,\right)$	$\left(\,-5\,\right)$	

Put these numbers in the squares   -5 -4 -3 -2 -1   1   2   3   4

Add across ⟶

Add down ↓

Sums are in ◯

	$2^1 - 4$		(-5)
$3 - 4^1$		Prime	(-1)
$4^1 - 3$		Square	(1)

(-5) (-9) (9)

Put these numbers in the squares   -5 -4 -3 -2 -1   1   2   3   4

Add across ⟶

Add down ↓

Sums are in ◯

		$\sqrt{64} \div -2$	**-3**
Odd		Odd	**-1**
	$\sqrt{64} \div 2$	$-9 \div 3$	**-1**
-2	**1**	**-4**	

Put these numbers in the squares   -5 -4 -3 -2 -1   1   2   3   4

Add across →

Add down ↓

Sums are in ◯

Even		Even
Odd		Odd
	$\sqrt{64} \div -2$	$-9 \div 3$

◯ **5**

◯ **2**

◯ **-12**

◯ **-2**   ◯ **-7**   ◯ **4**

Put these numbers in the squares   -5 -4 -3 -2 -1   1 2 3 4

Add across →

Add down ↓

Sums are in ◯

$-16 \div 8$			**-5**
Prime		$\sqrt[3]{8}$	**4**
		-3×3^0	**-4**
-4	**4**	**-5**	

Put these numbers in the squares   -5 -4 -3 -2 -1   1   2   3   4

Add across ⟶

Add down ↓

Sums are in ◯

		$2^4 \div 2^2$
	$\sqrt{1}$	$-25 \div 5$
	$-7 - -8$	

Circles (right): **3**, **-3**, **-5**

Circles (bottom): **-2**, **2**, **-5**

34

Put these numbers in the squares -5 -4 -3 -2 -1   1   2   3   4

Add across ⟶

Add down ↓

Sums are in ◯

$\sqrt{4}$			**-6**
	$\sqrt[3]{8}$	Prime	**2**
	2^2		**-1**
-9	**7**	**-3**	

35

Put these numbers in the squares   -5 -4 -3 -2 -1 1 2 3 4

Add across ⟶

Add down ↓

Sums are in ◯

	$2^3 - 2^2$	Even	
			(**2**)
			(**-2**)
	$\sqrt{25}$	Square	(**-5**)

(**-2**)   (**-4**)   (**1**)

Put these numbers in the squares   -5 -4 -3 -2 -1   1 2 3 4

Add across $\longrightarrow$

Add down $\downarrow$

Sums are in $\bigcirc$

$(30 \div -6)$			$\bigcirc$ -12
	Square	$-6 + 5$	$\bigcirc$ 1
3^1		$7^0 \times 8^0$	$\bigcirc$ 6
$\bigcirc$ -4	$\bigcirc$ 2	$\bigcirc$ -3	

Put these numbers in the squares -5 -4 -3 -2 -1 1 2 3 4

Add across →

Add down ↓

Sums are in ◯

		$7-8$	**-4**
Odd prime			**-5**
	$4^0 \times 4^1$	Prime	**4**
-3	**2**	**-4**	

Put these numbers in the squares   -5 -4 -3 -2 -1   1   2   3   4

Add across →

Add down ↓

Sums are in ◯

	$\sqrt{16}$	$-(2^2)$	**-5**
	Even prime	Odd prime	**6**
			-6
-5	**-4**	**4**	

Put these numbers in the squares   -5 -4 -3 -2 -1   1   2   3   4

Add across ⟶

Add down ↓

Sums are in ◯

	Square	Square	
			0
	Prime		**-2**
$\sqrt{1}$	Even prime		**-3**

-9   **6**   **-2**

40

Put these numbers in the squares   -5 -4 -3 -2 -1   1 2 3 4

Add across ⟶

Add down ↓

Sums are in ◯

		Odd
$5^0 \times$ -1	Even prime	
Odd		$\sqrt{4}$

◯ -6

◯ 5

◯ -4

◯ -1   ◯ -7   ◯ 3

41

Put these numbers in the squares   -5 -4 -3 -2 -1   1   2   3   4

Add across ⟶

Add down ↓

Sums are in ◯

1^{10}	$\sqrt{9}$	Odd	**1**
$10 \div -2$			**-8**
	Even	$-8 - -4$	**2**
-2	**0**	**-3**	

PRIME Center

MATHadazzles Volume 3

42

Put these numbers in the squares   -5 -4 -3 -2 -1   1   2   3   4

Add across ➝

Add down ↓

Sums are in ◯

$3^0 \times 3^1$	Even	
	$\sqrt{16}$	$\sqrt{16}$
$2 - 3$		

Sums (right, top to bottom): **0**, **1**, **-6**

Sums (bottom, left to right): **3**, **4**, **-12**

Put these numbers in the squares   -5 -4 -3 -2 -1   1   2   3   4

Add across ⟶

Add down ↓

Sums are in ◯

Odd prime			**-4**
		2^2	**-1**
	Square	-3×10^0	**0**
1	**-5**	**-1**	

44

Put these numbers in the squares -5 -4 -3 -2 -1 1 2 3 4

Add across ⟶

Add down ↓

Sums are in ◯

$3 - (2^2)$		
		$\sqrt{25}$
Even		Even

Circles right side: **-4**, **-4**, **3**

Circles bottom: **6**, **-9**, **-2**

Put these numbers in the squares   -5 -4 -3 -2 -1   1   2   3   4

Add across ⟶

Add down ↓

Sums are in ◯

	$\sqrt{16}$	10^0	⓪
Odd prime			⟨-3⟩
		$\sqrt{3^2}$	⟨-2⟩

⓪   ⟨-1⟩   ⟨-4⟩

Put these numbers in the squares   -5 -4 -3 -2 -1   1 2 3 4

Add across ⟶

Add down ↓

Sums are in ◯

-2 + -3			**-1**
		3× -1	**3**
-1^7		$\sqrt{16}$	**-7**
-2	**1**	**-4**	

47

Put these numbers in the squares   -5 -4 -3 -2 -1   1   2   3   4

Add across ⟶

Add down ↓

Sums are in ◯

-8 + 5			**-6**
		$2^3 \div 2^2$	**-5**
		$3^2 - 6$	**6**
-1	**-5**	**1**	

Put these numbers in the squares   -5 -4 -3 -2 -1   1   2   3   4

Add across ⟶

Add down ↓

Sums are in ◯

	2^2		**5**
$\sqrt{25}$			**-4**
$\sqrt{9}$	$-(2^2)$		**-6**
-5	**2**	**-2**	

Put these numbers in the squares   -5 -4 -3 -2 -1   1   2   3   4

Add across ⟶

Add down ↓

Sums are in ◯

Even prime			**-7**
Neither prime nor composite		Square	**2**
		Prime	**0**
2	**-9**	**2**	

Put these numbers in the squares   -5 -4 -3 -2 -1   1 2 3 4

Add across →

Add down ↓

Sums are in ◯

		Square	⎯ -1
3 − 7		Odd	-2
5 − 2×5	Factor of all numbers		-2

(-12) (-2) (9)

51

Put these numbers in the squares   -5 -4 -3 -2 -1   1 2 3 4

Add across ⟶

Add down ↓

Sums are in ◯

$5^0 \times -5$			**-12**
	$\sqrt{1}$	72^0	**-2**
		Even prime	**9**
-3	**-2**	**0**	

52

Put these numbers in the squares   -5 -4 -3 -2 -1   1   2   3   4

Add across ⟶

Add down ↓

Sums are in ◯

	2×-2		**-12**
Square	Prime		**8**
		$14 \div 2 - 5$	**-1**
-3	**-2**	**0**	

53

Put these numbers in the squares   -5 -4 -3 -2 -1   1   2   3   4

Add across ⟶

Add down ↓

Sums are in ◯

		$6 \div 2$	◯ **0**
Square	$2^2 \div -2$		◯ **-1**
2^0			◯ **-4**

◯ **0**   ◯ **-1**   ◯ **-4**

Put these numbers in the squares   -5 -4 -3 -2 -1   1   2   3   4

Add across $\longrightarrow$

Add down $\downarrow$

Sums are in $\bigcirc$

$-24 \div 6$		$7 - 8$	$\bigcirc$ -3
			$\bigcirc$ 3
	Prime	$\sqrt{25}$	$\bigcirc$ -5

$\bigcirc$ -3 $\bigcirc$ 3 $\bigcirc$ -5

Put these numbers in the squares   -5 -4 -3 -2 -1   1   2   3   4

Add across ➜

Add down ↓

Sums are in ◯

		Positive factor of 24
Positive factor of 24	$2^3 - 5^1$	
$\sqrt{25}$	24^0	

Row sums: **-1**, **3**, **-7**

Column sums: **-7**, **3**, **-1**

Put these numbers in the squares   -5 -4 -3 -2 -1   1 2 3 4

Add across ⟶

Add down ↓

Sums are in ◯

2×-1	$\sqrt{9}$	Negative	
			-3
Negative		Even prime	
			-4
	$\sqrt{9}$		
			2

-6   **-1**   **2**

Put these numbers in the squares   -5 -4 -3 -2 -1   1   2   3   4

Add across →

Add down ↓

Sums are in ◯

		Power of 2	
			-1
Prime	Prime		**4**
5× -1		-12 + 8	**-8**

-4   **0**   **-1**

Put these numbers in the squares   -5 -4 -3 -2 -1   1   2   3   4

Add across ➝

Add down ↓

Sums are in ◯

		$\sqrt{9}$	**-2**
	2^2	$\sqrt{25}$	**0**
$\sqrt{9}$	Positive		**-3**
-3	**2**	**-4**	

Put these numbers in the squares   -5 -4 -3 -2 -1   1   2   3   4

Add across ⟶

Add down ↓

Sums are in ◯

			0
$\sqrt{25}$	Positive	$\sqrt{1}$	**-2**
$9 \div -3$	Positive		**-3**
-5	**4**	**-4**	

60

Put these numbers in the squares   -5 -4 -3 -2 -1   1 2 3 4

Add across $\longrightarrow$

Add down $\downarrow$

Sums are in $\bigcirc$

$3^2 - 5$	Negative	Negative	$\bigcirc$ -3
	Odd prime		$\bigcirc$ -1
	Odd	$\sqrt{16}$	$\bigcirc$ -1

$\bigcirc$ 3 $\bigcirc$ 2 $\bigcirc$ -10

Put these numbers in the squares   -5 -4 -3 -2 -1   1   2   3   4

Add across $\longrightarrow$

Add down $\downarrow$

Sums are in ◯

Prime		$-18 \div 9$
	Prime	Odd
$\sqrt{16}$	$7 - 10$	

Circles on right: **-4**, **2**, **-3**

Circles on bottom: **-2**, **-6**, **3**

62

Put these numbers in the squares   -5 -4 -3 -2 -1   1   2   3   4

Add across $\longrightarrow$

Add down $\downarrow$

Sums are in $\bigcirc$

	Prime		
			-6
$\sqrt{7-3}$		$\sqrt{6-2}$	**-1**
Odd square	-2 × -2		**2**

-5   **6**   **-6**

63

Put these numbers in the squares   -5 -4 -3 -2 -1   1 2 3 4

Add across ⟶

Add down ↓

Sums are in ◯

Square		Prime
$6 \div 6$		
	$\sqrt{9}$	$-1 \times 16 \div 4$

◯ 5

◯ -6

◯ -4

◯ 2   ◯ -3   ◯ -4

Put these numbers in the squares   -5   -4   -3   -2   -1   1   2   3   4

Add across →

Add down ↓

Sums are in ◯

$\sqrt{16}$		$\sqrt{4}$	**-5**
	-12 + 7		**-4**
Prime		$\sqrt{16}$	**4**
-2	**-7**	**4**	

Put these numbers in the squares   -5 -4 -3 -2 -1   1 2 3 4

Add across ⟶

Add down ↓

Sums are in ◯

Prime	$\sqrt{16}$		**-4**
Composite	$\sqrt{4}$		**1**
$\sqrt{4}$	$\sqrt{1}$	$\sqrt{1}$	**-2**
5	**-3**	**-7**	

Put these numbers in the squares   -5 -4 -3 -2 -1   1 2 3 4

Add across ⟶

Add down ↓

Sums are in ◯

$\sqrt{4}$	$\sqrt{9}$	$\sqrt{16}$	**-5**
$\sqrt{16}$	$\sqrt{25}$	$\sqrt{1}$	**-2**
	Odd	$\sqrt{9}$	**2**
4	**-7**	**-2**	

Put these numbers in the squares   -5 -4 -3 -2 -1   1   2   3   4

Add across $\longrightarrow$

Add down $\downarrow$

Sums are in ◯

Odd		Prime
Even		
$4^2 \div -8$	Composite	

Circles (right side, top to bottom): **1**, **-7**, **1**

Circles (bottom, left to right): **1**, **-4**, **-2**

68

Put these numbers in the squares   -5 -4 -3 -2 -1   1 2 3 4

Add across ⟶

Add down ↓

Sums are in ◯

	Positive multiple of 2		**0**
Positive factor of 9	$-24 \div (3^2 - 3)$	Even	**1**
	Odd	$5 \times (6 - 7)$	**-6**

-2   **1**   **-4**

Put these numbers in the squares   -5 -4 -3 -2 -1   1   2   3   4

Add across ⟶

Add down ↓

Sums are in ◯

$15 - 8 \times 2$		Odd
	$2^3 \div 2^2$	
		Positive multiple of 3

Circles (right side, top to bottom): **-2**, **-7**, **4**

Circles (bottom, left to right): **-2**, **-3**, **0**

Put these numbers in the squares   -5   -4   -3   -2   -1   1   2   3   4

Add across   →

Add down   ↓

Sums are in   ◯

$\sqrt{25}$			**-12**
	Odd	Even prime	**9**
	$2^2 - 5$		**-2**
-3	**-2**	**0**	

Put these numbers in the squares   -5 -4 -3 -2 -1   1   2   3   4

Add across ⟶

Add down ↓

Sums are in ◯

Odd, not prime	$-(5^2) \div -5$	Positive multiple of 2	**-2**
			-6
Positive multiple of 3		Factor of 16	**3**
2	**-10**	**3**	

Put these numbers in the squares -5 -4 -3 -2 -1 1 2 3 4

Add across ⟶

Add down ↓

Sums are in ◯

		Positive factor of 5	**-4**
64 ÷ 8 ÷ 2	$\sqrt{25}$	Positive factor of 3	**2**
Even			**-3**
5	**-12**	**2**	

73

Put these numbers in the squares   -5 -4 -3 -2 -1   1   2   3   4

Add across →

Add down ↓

Sums are in ◯

		Factor of 5	④
	Odd		
			②
$(45-60) \div 3$		$\sqrt{16}$	
			⑦

(**4**)

(**-2**)

(**-7**)

(**-9**)   (**9**)   (**-5**)

Put these numbers in the squares   -5 -4 -3 -2 -1   1 2 3 4

Add across ⟶

Add down ↓

Sums are in ◯

	-2 + -2		◯ **-5**
		Factor of 7	◯ **0**
Odd		-1^3	◯ **0**
◯ **0**	◯ **-9**	◯ **4**	

Put these numbers in the squares   -5 -4 -3 -2 -1   1 2 3 4

Add across $\longrightarrow$

Add down $\downarrow$

Sums are in $\bigcirc$

		Prime factor of 9
$16 \div 4$		
	Even prime	$-10 + 6$

Sums across: -1, 1, -5

Sums down: 2, -4, -3

Put these numbers in the squares   -5 -4 -3 -2 -1   1   2   3   4

Add across →

Add down ↓

Sums are in ◯

			-9
	$2^3 \div (1 - 3)$	Composite	**2**
Odd prime		Odd	**2**
2	**-7**	**0**	

77

Put these numbers in the squares   -5 -4 -3   -2 -1   1   2   3   4

Add across →

Add down ↓

Sums are in ○

-2×2	Power of 2		**-1**
		Positive factor of 15	**-1**
$9 \div (-4 + 1)$	Power of 2		**-3**
-6	**1**	**0**	

Put these numbers in the squares   1   2   3   4   5   6   7   8   9

Add across ⟶

Add down ↓

Sums are in ◯

Odd	Even	$3^2 \times -1 + 5$	
	$-6 - {}^-1$		**-10**
Odd		Even	**6**

-1

◯ **2**   ◯ **-4**   ◯ **-3**

Answers

1.

1	2	3	(6)
4	-1	-2	(1)
-3	-4	-5	(-12)
(2)	(-3)	(-4)	

2.

1	-1	3	(3)
-2	4	-5	(-3)
-4	2	-3	(-5)
(-5)	(5)	(-5)	

3.

-1	-2	-3	(-6)
1	-4	-5	(-8)
4	3	2	(9)
(4)	(-3)	(-6)	

4.

-5	2	-1	(-4)
4	-4	3	(3)
-2	1	-3	(-4)
(-3)	(-1)	(-1)	

5.

4	3	2	(9)
-4	-3	-1	(-9)
-5	-1	1	(5)
(-5)	(-1)	(1)	

6.

-3	4	-4	(-3)
1	2	3	(6)
-1	-2	-5	(-8)
(-3)	(4)	(-6)	

7.

-5	1	-1	(-5)
2	-4	-2	(-4)
4	3	-3	(4)
(1)	(0)	(-6)	

8.

-5	-4	3	(-6)
-2	2	-3	(-3)
4	-1	1	(4)
(-3)	(-3)	(1)	

9.

-5	1	2	(-2)
4	-4	3	(3)
-1	-2	-3	(-6)
(-2)	(-5)	(2)	

10.

1	-5	3	(-1)
-1	-3	2	(-2)
-4	-2	4	(-2)
(-4)	(-10)	(9)	

11.

2	-4	1	(-1)
-2	4	-1	(1)
3	-3	-5	(-5)
(3)	(-3)	(-5)	

12.

2	-5	1	(-2)
3	-3	-2	(-2)
4	-4	-1	(-1)
(9)	(-12)	(-2)	

13.

3	-3	-4	(-4)
2	-2	-5	(-5)
1	-1	4	(4)
(6)	(-6)	(-5)	

14.

-1	-2	-3	(-6)
1	-5	-4	(-8)
2	3	4	(9)
(2)	(-4)	(-3)	

15.

-4	-1	-5	(-10)
-3	1	4	(2)
-2	2	3	(3)
(-9)	(2)	(2)	

16.

3	-2	-1	(0)
-3	2	-4	(-5)
4	-5	1	(0)
(4)	(-5)	(-4)	

17.

4	-2	1	(3)
-5	-1	-4	(-10)
2	-3	3	(2)
(1)	(-6)	(0)	

18.

-4	-1	-2	(-7)
-3	1	-5	(-7)
3	4	2	(9)
(-1)	(1)	(-5)	

19.

4	-5	1	(0)
-1	3	-2	(0)
-3	-4	2	(-5)
(0)	(-6)	(1)	

20.

4	-5	-3	(-4)
2	1	3	(6)
-2	-4	-1	(-7)
(4)	(-8)	(-1)	

21.

1	-2	3	(2)
-3	-5	-4	(-12)
4	-1	2	(5)
(2)	(-8)	(1)	

22.

1	-5	4	(0)
2	-1	-3	(-2)
3	-4	-2	(-3)
(6)	(-10)	(-1)	

23.

-3	-4	-5	(-12)
4	1	2	(7)
3	-1	-2	(0)
(4)	(-4)	(-5)	

24.

-3	-5	2	(-6)
1	-1	3	(3)
-4	-2	4	(-2)
(-6)	(-8)	(9)	

25.

-4	4	2	(2)
1	-3	3	(1)
-1	-5	-2	(-8)
(-4)	(-4)	(3)	

26.

-2	2	-5	(-5)
4	1	-3	(2)
3	-4	-1	(-2)
(5)	(-1)	(-9)	

27.

-4	-2	-1	(-7)
1	4	-3	(2)
3	2	-5	(0)
(0)	(4)	(-9)	

28.

1	-1	-5	(-5)
-4	2	-3	(-5)
4	-2	3	(5)
(1)	(-1)	(-5)	

29.

-5	-2	2	(-5)
-1	-3	3	(-1)
1	-4	4	(1)
(-5)	(-9)	(9)	

30.

-1	2	-4	(-3)
1	-5	3	(-1)
-2	4	-3	(-1)
(-2)	(1)	(-4)	

31.

2	-1	4	⑤
1	-2	3	②
-5	-4	-3	⑫

⟨-2⟩ ⟨-7⟩ ④

32.

-2	1	-4	⟨-5⟩
3	-1	2	④
-5	4	-3	⟨-4⟩

⟨-4⟩ ④ ⟨-5⟩

33.

-3	2	4	③
3	-1	-5	⟨-3⟩
-2	1	-4	⟨-5⟩

⟨-2⟩ ② ⟨-5⟩

34.

-2	1	-5	⟨-6⟩
-3	2	3	②
-4	4	-1	⟨-1⟩

⟨-9⟩ ⑦ ⟨-3⟩

35.

-4	4	2	②
3	-3	-2	⟨-2⟩
-1	-5	1	⟨-5⟩

⟨-2⟩ ⟨-4⟩ ①

36.

-5	-4	-3	⟨-12⟩
-2	4	-1	①
3	2	1	⑥

⟨-4⟩ ② ⟨-3⟩

37.

-4	1	-1
3	-3	-5
-2	4	2

(-4)
(-5)
(4)

(-3) (2) (-4)

38.

-5	-4	4
1	2	3
-1	-2	-3

(-5)
(6)
(-6)

(-5) (-4) (4)

39.

-5	1	4
-3	3	-2
-1	2	-4

(0)
(-2)
(-3)

(-9) (6) (-2)

40.

-3	-4	1
-1	2	4
3	-5	-2

(-6)
(5)
(-4)

(-1) (-7) (3)

41.

1	-3	3
-5	-1	-2
2	4	-4

(1)
(-8)
(2)

(-2) (0) (-3)

42.

3	2	-5
1	4	-4
-1	-2	-3

(0)
(1)
(-6)

(3) (4) (-12)

Answers

43.

3	-5	-2	(-4)
-4	-1	4	(-1)
2	1	-3	(0)
(1)	(-5)	(-1)	

44.

-1	-4	1	(-4)
3	-2	-5	(-4)
4	-3	2	(3)
(6)	(-9)	(-2)	

45.

-5	4	1	(0)
3	-4	-2	(-3)
2	-1	-3	(-2)
(0)	(-1)	(-4)	

46.

-5	1	3	(-1)
4	2	-3	(3)
-1	-2	-4	(-7)
(-2)	(1)	(-4)	

47.

-3	1	-4	(-6)
-2	-5	2	(-5)
4	-1	3	(6)
(-1)	(-5)	(1)	

48.

3	4	-2	(5)
-5	2	-1	(-4)
-3	-4	1	(-6)
(-5)	(2)	(-2)	

49.

2	-4	-5	(-7)
1	-3	4	(2)
-1	-2	3	(0)
(2)	(-9)	(2)	

50.

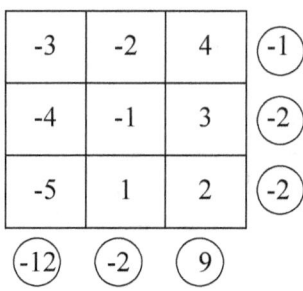

-3	-2	4	(-1)
-4	-1	3	(-2)
-5	1	2	(-2)
(-12)	(-2)	(9)	

51.

-5	-4	-3	(-12)
-2	-1	1	(-2)
4	3	2	(9)
(-3)	(-2)	(0)	

52.

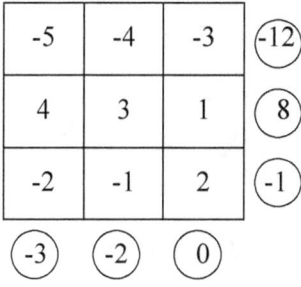

-5	-4	-3	(-12)
4	3	1	(8)
-2	-1	2	(-1)
(-3)	(-2)	(0)	

53.

-5	2	3	(0)
4	-2	-3	(-1)
1	-1	-4	(-4)
(0)	(-1)	(-4)	

54.

-4	2	-1	(-3)
4	-2	1	(3)
-3	3	-5	(-5)
(-3)	(3)	(-5)	

55.

-4	-1	4	(-1)
2	3	-2	(3)
-5	1	-3	(-7)

(-7) (3) (-1)

56.

-2	3	-4	(-3)
-5	-1	2	(-4)
1	-3	4	(2)

(-6) (-1) (2)

57.

-2	-3	4	(-1)
3	2	-1	(4)
-5	1	-4	(-8)

(-4) (0) (-1)

58.

-1	-4	3	(-2)
1	4	-5	(0)
-3	2	-2	(-3)

(-3) (2) (-4)

59.

3	-2	-1	(0)
-5	2	1	(-2)
-3	4	-4	(-3)

(-5) (4) (-4)

60.

4	-2	-5	(-3)
-3	3	-1	(-1)
2	1	-4	(-1)

(3) (2) (-10)

61.

3	-5	-2	(-4)
-1	2	1	(2)
-4	-3	4	(-3)
(-2)	(-6)	(3)	

62.

-4	3	-5	(-6)
-2	-1	2	(-1)
1	4	-3	(2)
(-5)	(6)	(-6)	

63.

4	-1	2	(5)
1	-5	-2	(-6)
-3	3	-4	(-4)
(2)	(-3)	(-4)	

64.

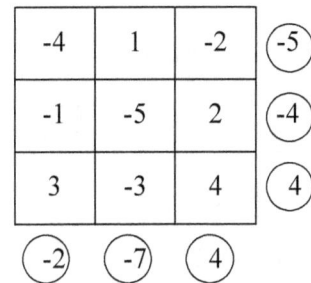

-4	1	-2	(-5)
-1	-5	2	(-4)
3	-3	4	(4)
(-2)	(-7)	(4)	

65.

3	-4	-3	(-4)
4	2	-5	(1)
-2	-1	1	(-2)
(5)	(-3)	(-7)	

66.

2	-3	-4	(-5)
4	-5	-1	(-2)
-2	1	3	(2)
(4)	(-7)	(-2)	

67.

1	-3	3	(1)
2	-5	-4	(-7)
-2	4	-1	(1)
(1)	(-4)	(-2)	

68.

-3	4	-1	(0)
3	-4	2	(1)
-2	1	-5	(-6)
(-2)	(1)	(-4)	

69.

-1	-2	1	(-2)
-5	2	-4	(-7)
4	-3	3	(4)
(-2)	(-3)	(0)	

70.

-5	-4	-3	(-12)
4	3	2	(9)
-2	-1	1	(-2)
(-3)	(-2)	(0)	

71.

1	-5	2	(-2)
-2	-1	-3	(-6)
3	-4	4	(3)
(2)	(-10)	(3)	

72.

-1	-4	1	(-4)
4	-5	3	(2)
2	-3	-2	(-3)
(5)	(-12)	(2)	

73.

-1	4	1	(4)
-3	3	-2	(-2)
-5	2	-4	(-7)
(-9)	(9)	(-5)	

74.

-5	-4	4	(-5)
2	-3	1	(0)
3	-2	-1	(0)
(0)	(-9)	(4)	

75.

1	-5	3	(-1)
4	-1	-2	(1)
-3	2	-4	(-5)
(2)	(-4)	(-3)	

76.

-3	-1	-5	(-9)
2	-4	4	(2)
3	-2	1	(2)
(2)	(-7)	(0)	

77.

-4	4	-1	(-1)
1	-5	3	(-1)
-3	2	-2	(-3)
(-6)	(1)	(0)	

78.

1	2	-4	(-1)
-2	-5	-3	(-10)
3	-1	4	(6)
(2)	(-4)	(-3)	

www.ingramcontent.com/pod-product-compliance
Lightning Source LLC
Chambersburg PA
CBHW060402190526
45169CB00002B/717